STEM CELLS
IT'S NOT SCIENCE FICTION... IT'S SCIENCE IN ACTION

Dra. Adriana Gudiño

BARKERBOOKS

■ BARKERBOOKS

STEM CELLS IT'S NOT SCIENCE FICTION... IT'S SCIENCE IN ACTION
All rights reserved. © 2023, by ADRIANA PAULINA GUDIÑO REYES

Copyediting & Proofreading: Yeni Rodríguez| BARKER BOOKS®
Book Cover Design: Sara Risk| BARKER BOOKS®
Book Interior Layout: Sara Risk | BARKER BOOKS®
Cover photo: Shutterstock/ENVato | BARKER BOOKS®.

First Edition. Published by BARKER BOOKS®
Paperback ISBN | 979-8-89204-713-5
Hardcover ISBN | 979-8-89204-714-2
eBook ISBN | 979-8-89204-712-8

Library of Congress Copyrights Control Number: 1-13454762261

No part of this publication may be reproduced, distributed, or transmitted in any form or by any means, including photocopying, recording, or other electronic or mechanical methods, without the prior written permission of the publisher, except in the case of brief quotations embodied in critical reviews and certain other noncommercial uses permitted by copyright law. For permission requests, write to the publisher, addressed "Attention: Permissions Coordinator," at the e-mail address below. The information, opinion, analysis and content of this book is the author's responsibility and it does not represent the point of view of the editorial company, BARKER BOOKS®.

Printed in the United States.
BARKER BOOKS® and their affiliates are an imprint and registered trademark of Barker Publishing, LLC

Barker Publishing, LLC
Los Angeles, California
https://barkerbooks.com
publishing@barkerbooks.com

Dedication

This book is dedicated to my parents, who always motivated me to achieve my goals. To my dad, who is in heaven, who one day before his death decided to put himself in my hands with my treatments, but it was too late, we relied on conventional medicine during his last moments of illness.

Introduction

The topic of stem cells is as important as it is controversial, undoubtedly because of all the lies and myths that have been spread about this science, either due to ignorance or misconception. However, there are a few scientists who are faithful to this regenerative science. As a professional I have applied it for 16 years and I have obtained excellent results, with full knowledge of the cause, because nothing in this branch of science is random, everything is specifically focused. I work with scientific tables that for a long time I have shared with Ricardo Rangel, a certified biologist. We share a wonderful synergy that finds its peak within a laboratory with state-of-the-art equipment and high quality standards, the protocol, first and foremost, contains human warmth and empathy.

As a faithful believer in this science, I am fully involved in the whole process. I myself have witnessed all the skilled work that, behind pipettes and flasks, is carried out around the development of stem cells: their extraction, their cultivation, their reproduction and their growth. This is an amazing process, watching and admiring how a cell lives and grows, which in the end will be vital in the health of women and men. My daily work includes reviewing and overseeing various clinical cases at every stage.

Although this therapy is avant-garde in ultra-developed countries such as Switzerland, Mexico is still in "nappies", far behind in the requirements, evolution and progress in this field. At present, as if it were a panacea or a modern issue, there are a wealth of clinics scattered throughout the country, boasting of curing any disease. They claim to be the pioneers in this field, when the reality is very different. These are run by companies that only seek profit and financial gain, not caring about the pain, anguish and desperation that sick people suffer; they do not treat people for the sake of ethics or neighbourly love. There are only a few companies that offer stem cell treatment, thanks to high-ranking scientific research and true cases of success.

MER is a specialised clinic with high standards of health, efficiency, protocol and scientific etiquette in stem cell treatments. It is here that my work converges with the timely assistance of a specialised laboratory. My vast experience, together with my knowledge, have given me status and recognition in the regenerative field. I have detailed information on the subject of diseases: what they are, how they operate and how to eradicate them with treatment. That is true regenerative medicine, without a doubt. I struggle every day with the limits imposed on my renowned work by those profitting from people's pain and despair.

In fact, I have been faithful to this science for a long time, expanding my preparations day by day by studying, revising, creating, generating and contributing the extra work that is needed, because I believe in this wonderful knowledge. I know that if it is included in all teaching schedules, soon Mexico's worrying health statistics will improve, and there will be many more satisfied patients, who are happy, healthy and living dignified lives, like a patient of mine whose sight was recovered after prolonged blindness by a personalised treatment. I could cite more cases of success, but what matters is what can be proven with evidence, not with just words or profit, much less lies or attack.

My presence in this field is rock solid, my convictions are strong and I am motivated by genuine altruism and social empathy. I am building a mountain of success one grain of sand at a time, with each grateful patient sleeping well, eating well, and living a dignified life with their loved ones and without pain.

I believe in regenerative medicine, but I believe more in myself, in my capacity as a professional. I am sure that if many others began to believe in themselves, they too would also soon have good results.

Finally, I must point out with sadness that the outcomes, the health, the well-being that regenerative medicine can achieve for numerous ailments, is currently very restricted by so much controversy and criticism, much of it baseless and unfounded. This is largely due to the powerful pharmaceutical companies that restrict our work, creating false expectations about health, medicine and its side effects. Stem cells treatments, according to their simple philosophy, would likely take a big slice of their cake...

Chapter 1
Stem cells, a myth-busting science that creates hope

You get a wound and in a few days it heals. You get an infection from eating street food, your immune system becomes alert and arrives with a whole natural army of T-cells, fighting bacteria, viruses, fungi and other germs. In other words, your body's sacred bunker is designed to heal itself; it has an innate self-healing capacity, creating an internal balancing process called homeostasis. A human body is a powerful self-healing organism. It does it automatically, naturally, without conscious participation, while you sleep, eat or work - amazing, isn't it?

Your body is marvellous and extraordinary, the most perfect machine that exists (and at the same time the most vulnerable), with its efficient internal organs, each one with its own task, but at the same time collaborating as a team. It is a complex mathematical and engineering marvel. For example, the most powerful computer, without a doubt, is the brain. Just to open a can of chili peppers, it has to send millions of nerve signals per second to the hand, while simultaneously co-ordinating a multitude of diverse bodily functions. These stimuli are transmitted at a speed of more than 400 km/h.

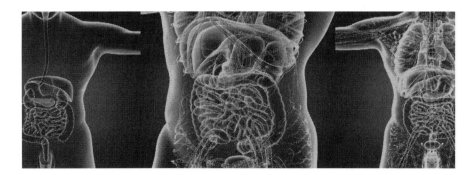

The brain is able to do this because it processes information in a unique and concrete way. It divides the work among its 100 trillion cells, and each of these in turn communicates with ten thousand others through a network of electrical connections. And it doesn't end there, behind everything you do in your daily life there is an army of microscopic machines, responsible not only for brain actions, but for all bodily processes: 50 trillion cells. Your body IS cells! Inside each one are hundreds of power stations, protein factories, recycling plants, and transportation highways that carry, bring, support, nourish, nurture, and give life. And right inside this cellular chest are stem cells.

What comes to mind when you hear or read "stem cells"?

The controversial stem cells are those that are in an early or initial phase of development, in which stage they keep all that potential to multiply and become different and interesting types of cells within the organism. It should be noted that most adult human cells have a specific purpose that cannot be modified. However, under certain conditions, they can form any of the tissues and organs of the human body, giving it new life - incredible, isn't it?

In other words, like a working machine the noble objective of these microscopic elements is to trace, identify and diagnose in order to repair whatever needs to be repaired, be it the repair system, engineering and biology of tissues, organs, or cartilage.

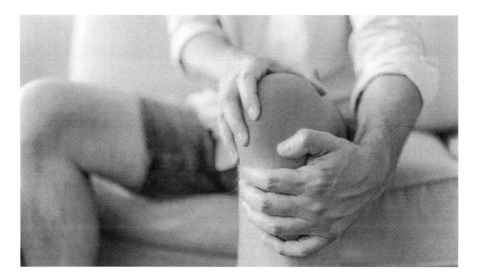

This sounds like lines from a book by Stephen Hawking, a script for a futuristic movie or the documentary that tells the story of a child who relieved his pain thanks to a stem cell treatment. In any case, the reality today surpasses fiction and silences speculation, since stem cells also surpass utopia, beliefs, the controversy that once arose and, above all, the ill-intentioned myths such as that stem cell usage:

- Is about cloning
- Is about experimentation with embryos
- Generates undesirable side effects, such as cancer.

Cancer Testimony

In our clinic, MER, we have more than 16 years' experience treating active cancers and their sequelae, since these substances are immunomodulators, they raise the patientьs defenses so that he/she can defend him/herself from the malignant process.

Much of the controversy and dispute is the result of incorrect and even misleading information that hinders the progress of this regenerative medicine, as extraordinary as it is effective. The propagation of unscientific rumours has caused damage to research and has somewhat stagnated the use of this branch of treatment in its application in specialised laboratories and clinics that provide health care.

Currently this science is available in the here and now, presented as the hope of life and relief from so many ailments that until now have been classified as "incurable", degenerative and disabling in many cases, such as cancer and diabetes, even autoimmune diseases that plague the planet, pain conditions that so far traditional or conventional medicine have not been able to address. It is not science fiction, it is science in action, today's medicine for the future controlled by strict ethical standards.

Cell research took its first steps in the mid-19th century, when physicians and scientists found that cells were the building blocks of life. By the 1950s, physicians had performed nearly 200 allogeneic stem cell transplants in humans. By the late 1960s, they were successfully used in bone marrow transplants to treat immune deficiencies, anaemia and leukaemia. As you can see, this is not a new topic or one that has been pulled out of a hat, but one that has earned not only respect and recognition, but also accreditation and approval from top health institutions, since they see in this science a panacea that to this day provides quality of life through healing and well-being in general.

Chapter 2
Stem cells: scanning, detecting and repairing

Today, regenerative medicine is leaving fiction behind to become a reality that brings hope to so many who suffer the pain of chronic degenerative diseases and autoimmune diseases. The future is already here thanks to the promising advances of this type of cutting-edge medicine supported in three different spheres:

- Genetic Engineering.
- Tissue Engineering
- Cell Therapy.

The current progress of the latter has been astonishing, also called the third medical revolution, after the vaccines and antibiotics of the 20th century. The good use of stem cells allows, broadly speaking, the repair of damaged tissues in the organism, but also a wide range of specific functions. This gives rise to uncertainty and certain doubts that hinder treatment.

These are living substances obtained in a specialised laboratory from a human placenta previously studied and analysed during the gestation period of a pregnant patient. After a whole process, a healthy, strong and pathogen-free "product" is obtained. Subsequently, biologists and specialised chemists cultivate it in a laboratory, where these stem cells are reproduced. It should be noted that these stem cells, unlike exosomes, take a little longer to act because their molecular weight is greater. However, when they do, it is in a peculiar way because of their effectiveness, they have a great regenerative capacity.

Did you know that...?

- When a stem cell is cultured, it secretes exosomes, cellular proteins packed with nutrients. As nanoparticles, they travel at the speed of light through the microcirculation, detecting inflammation in the body faster.
- Exosomes + stem cells = healing synergy.
- Regeneration = repair and restoration of damaged tissues. Stem cells are regenerative, high biotechnology. They go straight to the root of the problem or ailment. They are not drugs or painkillers that only numb the symptom, they scan and repair organs, tissues and bones, without leaving sequelae, acting quickly and effectively.

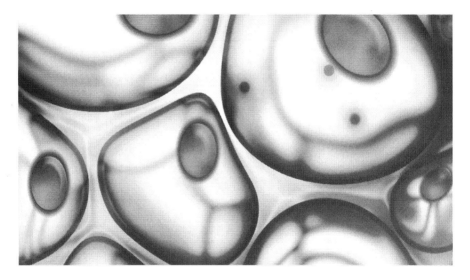

Placenta cells from another person inside my organism?

That is so; however, doubt and speculation translate into truth after a whole process of research and treatment. The human placenta is a virgin tissue, pure, it does not store antigenic load in its cellular wall, moreover, it does not generate adverse or allergic reactions in a patient/recipient. It should be noted that each patient is unique and functions, thinks and feels very differently, hence the only side effect is fatigue. In addition, achieving these substances entails a rigorous and arduous procedure, covered by strict health, safety, professionalism, teamwork, efficiency, ethics and high-quality standards. These elements include::

- An *ex-profeso* laboratory, very professional in its kind.
- Doctors, biologists and chemists with expertise in the field. Together, they analyse, safeguard, ultra-disinfect, culture and process with proven, state-of-the-art reagents.

That is to say, this regenerative medicine is scientifically produced by up to the minute knowledge, handcrafted, detailed and closely monitored so that the results in the patients are evident and supported. Research is the basis of any treatment..

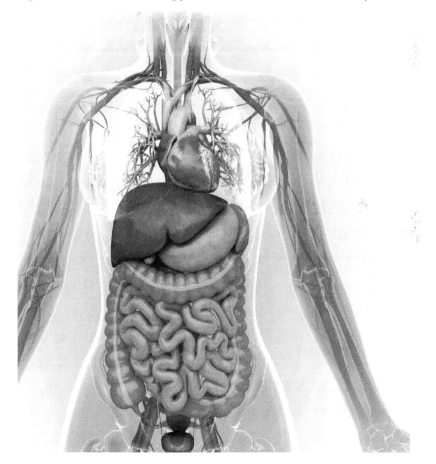

Preparing the patient

Stem cells are noble, kind agents, living substances, they are the raw material of the body, the basis of life. In addition to generating tissues and organs, along the way they create new cells when those that already exist end their life cycle. From them are generated all those millions of cells that fill our body (daughter cells). In order to carry out efficient therapeutic work, it is necessary to have the most favourable environment possible. Prior to treatment, the patient undergoes a "detoxification" that includes:

- Adding fruits, legumes, fibre and vegetables to the diet; the latter have a high anti-inflammatory potential for the cells in the organism.
- Enough water to create strong hydrated tissues, to enable the stem cells to grow, develop and work on their healing treatment properly.

The plan is for them to find the most favourable environment possible for them to facilitate their regenerative work.

Did you know that...?

A stem cell has the innate capacity to generate mature cells that are more specialised than itself, in order to form tissues and organs and eventually repair them. If it were not for them, a small injury would be fatal.

Chapter 3
How do stem cells work??

The therapeutic work, broadly speaking, is to go, detect and repair damaged areas within the organism, albeit tissue, an organ, cartilage, the liver, a lung. The curative work can be somewhat uncomfortable, and take from 24 to 48 hours, so the patient must rest.

It is usually scheduled again one month later. This is enough time for the stem cells and exosomes to produce excellent results. It is worth emphasising the prompt action of these nanoparticles, almost immediately in terms of compensating blood pressure or reducing the symptoms of neuropathy or diabetic retinopathy. This is biological medicine that in a short period of time provides well-being, since it generates a process called angiogenesis; that is to say, new blood vessels vascularise the tissues again, and in this condition they repair and revive an organ that has been given up on by conventional science.

As a specialist in regenerative medicine, I take a timely clinical history of how the patient arrives and leaves after stem cell and exosome treatment. This is the evidence that demonstrates the veracity and effectiveness of the treatment. The results are justified in order to eradicate any disbelief and mistrust, avoiding ethical doubts, results that are demonstrated through the testimonies of the patients themselves when they witness the moment when their liver enzymes, sugar levels or cholesterol reduction decrease.

Side effects?

As a concept, there are none. If anything, there is a slight fatigue as the immune system fights against these substances, trying to recognise that what enters the body are not pathogens. There is also what we call a healing crisis, which is the body's natural reaction to eliminate the damage and regain the lost balance, and along the way it makes peace with the immune system by offering immunomodulation so that it can operate in balance.

And to reinforce the task or as a preventive measure, the treatment is repeated the following year. It must be said that each patient has a unique clinical history because of the diseases and hereditary factors he or she carries, which is why it is a unique treatment, with precise doses of stem cells.

Regenerative medicine replaces, repairs and offers health when many doctors have already discarded it. It is a whole healing process that begins with the innate intelligence of these primordial units to travel freely throughout the body, recognise, scan, detect and repair internal damage, whether it is a pancreas that does not secrete enough insulin, the deficient cartilage of a knee, or the function of that kidney that would be replaced.

- It scans and repairs damage, including some hormonal deficiency in the brain, to replace and place new substances.
- In their journey they clean or purify the cellular oxidative stress, the excess of free radicals that oxidise the cells so that they do not do their work.
- Insomnia? Perhaps there is too much cortisol, the stress hormone. Stem cells regulate and create new connections

We will never cease to be amazed by this science - the biological treasure of the 21st century. The future is already here, the eradication of diseases that used to cause deaths every year will no longer be so with the correct use of this medicine.

Did you know that...?

Cellular ageing begins between 35 and 40 years of age. One of the dreams of humanity, almost impossible being utopia, is a future without diseases and to stop the hands of the biological clock that makes us mortal, perishable - the tireless search for eternal youth. Growing old is an obligatory stage of life, that is why it is so logical to understand and deduce that we all reach this stage, and that in doing so, so do the millions of cells, those tiny inhabitants inside, outside, in the middle and in all the co-ordinates of our body.

It is a simple biological fact that can be reversed to a certain extent if we lead a good life, without excesses, away from bad habits such as alcohol, cigarettes, bad diet and late nights. The mechanisms of cellular ageing start from the very moment we are born. Cells live, reproduce, divide, multiply and perform their assigned functions, like precise orders given to a soldier.

And when it has completed its life cycle, it is replaced by a new, healthy cell, genetically equal to the mother cell. But after years of hard work, they age, their functions slow to nothing, until they die and cannot be replaced. Take care of them as much as possible, fortunately you do not need a degree in molecular biology to make sure that they are strong and healthy. Eat healthily, sleep well, exercise, don't smoke or drink excessively. They will thank you by giving you health and well-being.

Chapter 4
10 curiosities about stem cells

1. Who takes care of your cells? Proteins help your body generate shiny new spheres and repair old ones.
2. 2They detect enough potential in their DNA to efficiently treat a wide spectrum of diseases such as diabetes, kidney failure and heart disorders, and also neurodegenerative diseases such as Parkinson's and Alzheimer's, among many others.

3. In a simple way, stem cells divide giving rise to two daughter cells, the heirs of the human system. One of them, perhaps the chosen one, is identical to the progenitor, that is to say, it has the same traits, the way it moves and behaves in the organism, also the same properties and can divide again if it needs to (cell self-renewal). The other, no less important for the mother, has the magnificent capacity to become a specialised daughter, which carries out a very specific task, in such a way that it adapts to various types of tissues to regenerate or repair them in case they are damaged.

4. Reliable data shows that today stem cells are used in more than 80 diseases such as:
   - Diabetes.
   - Renal insufficiency.
   - Arterial hypertension.
   - Orthopaedic injuries.
   - Cardiovascular disease.
   - Erectile dysfunction.
   - Some serious disorders such as Parkinson's, Alzheimer's and multiple sclerosis, where it is highly effective and without adverse effects.
   - Research in the field continues its course with good news, hence in the near future, this regenerative medicine will be the hope for life and dignified health that we all deserve.
5. A stem cell has the exact intelligence to divide indefinitely: 2, 3, 4, 6, 100, 1000... And thus differentiate into different types, each one with a different shape, colour, texture, volume, diameter, work and task to do.

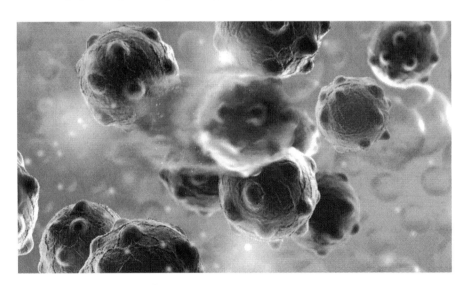

6. According to their origin, these are classified as embryonic (extraembryonic) and adult. The former are those present in the early stages of the embryo, from which all cell types arise (they create systems, tissues and organs). The latter are found in adult organisms. On the other hand, there

is another classification that depends on the capacity, strength and function performed by each one; thus there are totipotent, pluripotent, multipotent, unipotent and oligopotent cells.

7. It is in a highly specialised laboratory, under strict safety and ethical standards, where healthy cells can be manipulated or generated to replace old or diseased ones (regenerative medicine), with the noble purpose of regenerating and healing damaged tissues and organs.

8. How old are you right now? Whatever it is, it is not the age of some parts of your body. Your tissues, organs and cells show very different ages. Many cells will not even have hours of existence, they are renewed over and over again. But when an organism is damaged and creates bad living habits, this renewal is slow or dysfunctional, hence the urgency to generate a healthy lifestyle.

9. Stem cells also treat autoimmune diseases with good results, such as multiple sclerosis, type 1 diabetes, rheumatoid arthritis and others, as there are more than 80 types. An autoimmune disease appears when your immune system attacks your healthy cells of organs and tissues, due to confusion, disorientation or error. It does not recognize them, it "sees" them as foreign agents, as germs. But in this case, stem cells, entering through the bloodstream, stop this damage.

10. A near future without disease? We are already living it with the research and accurate treatment of stem cells. This is a reality that adds hope in a world where epidemics ravage our days. Thousands of scientists are now focusing their eyes and knowledge on these life-giving microspheres. More healing tools are expected from stem cells. The medicine of the future is already here and must be used with caution, but with this humanitarian vision. They make this world, infested with all kinds of diseases, a more bearable one, among so many forgotten patients. Doctors should put aside their ego to give way to help the most needy, who survive in the most adverse circumstances. This is part of my stem cell treatment decalogue..

Chapter 5
Autoimmune diseases and how to treat them

An autoimmune condition arises when your immune system attacks your own body, healthy cells and tissues by mistake, perhaps out of disorientation, confusion or because it believes there is danger inside.

Lupus is one of many autoimmune diseases, and is very serious and painful for those who suffer from it, as it damages joints, kidneys, lungs, liver, heart, skin, etc. It hurts, ruins and generates in the patient a feeling of helplessness and suffering, both physical and emotional,

hence the wonderful importance of exosomes and stem cells in their treatment and potential cure. They travel through the blood and along the way they scan, detect and restore a faulty, deficient immune system and stop the process of self-destruction. They also help repair those organs that were injured, such as the liver and kidneys.

In Hashimoto's thyroiditis, stem cells restore the system, stop the deterioration and begin to activate thyroid function. The process is almost the same in rheumatoid arthritis; they enter, circulate and detect the affected joints, interrupt and at the same time heal the damage, remove the pain and reduce inflammation. The results are visible in a month or sooner depending on each patient.

This is an integral treatment that includes a correct diet/nutrition. Undoubtedly, this therapy, until recently controversial and condemned in medical circles, is today the hope of so many sick people, with feelings of concern and helplessness before the "incurable" disease. It is a noble, effective procedure, without patent medicines that only produce sequelae.

When my own body attacks me... and generates autoimmune diseases

In the medical encyclopedia there are dozens of autoimmune diseases, mostly chronic or severe, almost incurable due to the aggressive pathology they contain and with a high risk of mortality. Examples include lupus, fibromyalgia, multiple sclerosis, amyotrophic lateral sclerosis, Sjogren's syndrome, vitiligo, and so on. Hence, stem cell treatment must be specific, punctual and of the correct dose to deal with the damage, restricting it until it is eradicated without side effects.

Stem cell treatment is the valuable effectiveness of a far-reaching scientific procedure, a therapeutic potential within regenerative medicine that restores the order between organs and integral well-being for a dignified life, a human right enshrined in the constitution, as a good and a virtue for women, men and children.

Dare to change your life!

The benefit of the doubt is in you, leave the beliefs, myths and rumours of cloning and cancer, dare to change, to be a better person when health does its work inside you! Do not get used to the pain, do not normalise it in your daily life, do not resign yourself or be afraid to take this treatment.

Fear is a natural instinct that keeps us alert, the problem is that for many this emotion functions as the starting point in decision making, and this fear includes questions like: "What if it doesn't work? What if I get worse?" Many stay on this step because they begin to look at obstacles that translate into, "I think I won't be able to".

Gag that internal terrorist that activates itself to shout, "You can't do it!" by reminding it of the times you dared to do something - and you succeeded. Most of us, when faced with this type of scenario, turn on autopilot or react instinctively, and it's even normal, since we intend to change something that has been there for long years, normalising suffering and illness. "I can't", two words with a determining mental effect, which the subconscious picks up on, so ignore it so that it does not sabotage you!

Change it to, "Yes I can and I will try with stem cells!" Focus on your strengths (not your limitations), what makes you strong physically, emotionally and spiritually. The one thing you can be one hundred percent sure of is that if you don't try, if you don't change your medicine, you may later get a bill for more pain than you feel today.

Did you know that...?

Your immune system is a natural army of T-cells that keep viruses, bacteria, fungi and other germs at bay. Sleepless nights, stress, poor diet, drug abuse, pollution, and sudden changes in temperature among other things, diminish this defense so that any disease can harm you. Strengthen it with stem cells!

Chapter 6
Between your body and emotions

Your body is the temple of nature and of the divine spirit. Keep it healthy; respect it; study it. Grant it its rights.
Henry Frédérick Amiel

Hello, I am your body! Do you want to be healthy, to no longer wake up at dawn with that critical pain? There are many ways to achieve this end, some of which are common sense:

- Balanced diet.
- Exercise and recurrent activities.
- Avoid bad vices such as alcohol, cigarettes and addictive substances.
- Positive thoughts, get away from stress, resentment, indifference to yourself, your family and all those around you.

And, above all, it starts by paying focused attention to your body, listening to it, attending to its alarms and red lights. How many times have you felt exhausted, stressed, with a sharp pain that sticks deep in your stomach - and you continue with what you are doing? "It will pass, it's normal, I'll take a pill, it's more important to finish this work.".

Be apathetic, normalise the pain and put a band-aid on it, prioritise the daily routine. Surely since you wake up, your brain is running, thinking about pending work, chores, family, children, partner, friends, projects. Chaos circulates through your cloud storage, a hotbed of neurons, impulses that at the moment do not know whether to connect or jump. Your mind is still working even on Saturdays, Sundays and holidays.

That's what most people do, and time consequently casts unfavourable forecasts not only for chronic stress, but also for body bills, inside and out. Instead of paying attention to that internal radar, you keep going with coffee, pills, late nights and bad eating habits. Many people today tend to live under stress, which is also normalised as a negative, critical, complaining mood. The body cries out that it has run out of energy, out of resources, that it hurts below, above, in the middle and all over the body. As human beings we are separate, dissociated. The body and the mind play in different teams and we feel that we are pulled without knowing what we should listen to: either to the mind that advises you to go on (but at the same time to stop) or to the body that refuses to do so.

Many of us choose to listen to the mind. We do not listen to our body, we ignore it, even though there is already water everywhere. You must learn to listen to your body, pay attention to it even if it is just a buzzing in your ears. Surely at this moment you are reading, something hurts you, oppresses you, makes you uncomfortable, irritates you, buzzes or makes you dizzy, because it is shouting at you. Do you listen to it, do you change your routine that includes a bad diet to pay attention to priorities? Probably not and you leave it for tomorrow like the course or the exercise of five years ago, and obviously the symptom turns into a serious ailment from which often there is no way back - just wait for the outcome. There are many who come to the office with a serious case of diabetes or arterial hypertension and even renal insufficiency that prostrates them with pain, worry and uncertainty. Maybe it has already happened to you, if it hasn't, is that what you want? Listening is the key, prevention is what follows.

Just as you listen to the complaints of your neighbours, listen to the complaints of your body. Listen and pay attention to the first symptom. Who better than your body to point out a pain, a laugh or a poorly channeled emotion? Get in tune with the areas that feel relaxed and with those that are in discomfort. Listen to it, feel it, touch it and listen to it. Don't underestimate what is screaming at you, and act on it immediately.

- Don't put off until tomorrow what you can fix today.
- Don't put off until tomorrow that consultation that you have already postponed several times.
- Do not wait until tomorrow to start that diet, that exercise.

- Do not wait until tomorrow to live today without any pain or illness.

What hurts right now? Maybe you have noticed that twinge that sinks like an arrow under the ribs on your right side or the sharp pain in your knees because you were told their cartilage is gone. Your body is screaming, asking for help, "Please stop eating so much sugar, fat, flour, salt, spicy, processed foods!". And in this complaint, many of your organs beg because they want to stay by your side, not to be removed because of you. Don't ignore this SOS, heed this call when, right after finishing that delicious hamburger, you feel that pain under the last rib on the right side.

Stop being patient and turn that energy into a plan of supreme wellness for your body and emotions. Remember, there is no cause without effect. Not believing in this maxim is like denying the theory of gravity or living with bad daily habits, and thinking that the body and emotions will not write a bill with many zeros.

Fear? Do not listen to fear, because it is a bad advisor that can turn into panic. Rational fear saves your life, it tells you where to go, what to do, what medical decision to make. Irrational fear first makes you suspect, conjecture, suppose, anticipate, paralyse and imagine what might happen if your kidneys stop working suddenly or bit by bit, perhaps with dialysis.

When faced with the new, unknown and threatening, it is natural to feel fear; it is a natural instinct that keeps us alert. The problem is that for many this emotion functions as the starting point in decision making - deciding not to take that health option because you were told it is unnatural. How you occupy your thoughts greatly influences your actions. You must focus what you think on what you want to happen. In other words, direct that energy toward a goal - to protect and heal.

Chapter 7
We are mind/energy

Ladakh, India, 2023

We are in an earthly world where physical pain tells you that something is happening in one of your internal organs. Do you take care of your body, or at least try to? We always deal with physical discomfort, but what about your mind? Did you know that it requires some continuous cleaning, since day by day you deal with eventualities that can also make it sick if you do not perform the much needed mental cleansing.

We are mind

We do not have the slightest idea of the great power that dwells within us, it is enough to know that daily stress generates cortisol and this hormone damages the body cells if we do not do something to channel it, for example:

- Meditate - empty all those daily thoughts that bother you, heal it and make a daily *reset*. Breathing helps a lot to control stress levels. Take three deep breaths when you feel very stressed.
- Practice yoga or have a relaxing massage.

- Stay in contact with nature as much as possible, it relaxes the mind.
- Live with the animal you like the most, this reduces stress levels.
- Undertake a physical activity that produces natural hormones, such as dopamine and serotonin, which counteract the damage caused by stress.
- Stay close and at peace with your loved ones, this peace of mind lowers stress levels.
- Read a book that transports you to the world of adventure and knowledge.

When you nourish your body in a healthy way, hydrate it constantly and provide it with adequate sleep, mental cleansing can flow much better. Don't forget to take a few minutes a day to clear it. You will notice that by doing this daily cleansing, you will eliminate chronic pains and you will be able to reduce medications. For these reasons it can be said that the mind is so powerful that it can cure you of an illness, but be careful! It can also make you sick, that's why we have to be very careful and precise with what dwells in our mind.

The objective is that you connect your being (mind-body-spirit) and you can converse with the universe; you have to be attentive to the daily signals that your own consciousness gives you, if you reach this point, you only have to ask and it will be given to you, but be careful... think well about what you want before asking the universe, it will surprise you if you don't do it correctly.

Be honest with what you want, don't only converse with the universe, you can also do it with yourself, putting internal limits, discipline in your thoughts, without forgetting to reward yourself for your small or big achievements. Be your own best friend - the first thing you have to do is to know yourself and like yourself, love yourself deeply so that everything you do you enjoy, you never feel alone and have the best motivation: "yourself". Don't forget, the answers are inside you!

Meditation

Chapter 8
Do you feel guilty about being sick?

For many suffering from diabetes, kidney failure, even cancer, or being overweight is to carry a great sense of guilt: for not taking care of themselves, not going to the doctor, not keeping good habits. They feel guilty, believing that they are a burden to their family. The feeling of guilt is like that internal judge that many of us carry inside, that reproaches and conditions us for some real or imaginary fault.

There are so many "faults" for which remorse is felt and makes us look in the rearview mirror, and from there to the most intense depression, bitterness, anguish for not getting any relief by knocking on doors, going to therapists, receiving "placebos"... with no sign of improvement. Surely you are already desperate, because the disease, on the contrary, grows.

The feeling of guilt is not bad in itself, it makes us aware that we have acted badly in something which facilitates the attempt to repair it, consider other health options beyond the conventional and good lifestyle habits, to love your body above all things; it helps us to learn from our mistakes and grow. But it is very different to feel guilt as a normal or empathic instinct than to blame oneself mercilessly for all the pain that keeps us in our bed, by action or omission or to beat oneself up with a hurtful internal dialogue that undermines self-love: "for being stupid, for being stubborn, for listening to the rumour".

The root of this feeling could be found in one's education, full of reproaches where punishment was valued more than forgiveness. It could also be in the expectations and/or demands of perfection that others put on our shoulders, and to what our social environment believes is expected (marriage, children, work, achievements, physical and emotional health to 100). When we don't fit those moulds, frustration sets in... and later, guilt.

Catharsis. Write a guilt diary (habits, circumstances, carelessness, listening to others, etc., you will know your story). Include people, write down all those burdens you experience, scenarios where you feel responsible for something that happened: binge eating, sedentary lifestyle, junk food. Examine your conscience and see if these really have the value you place on them, then you will deepen the level of guilt to diminish its emotional effects.

Acceptance. How many qualities do you consider good about yourself, but distasteful to others? If you seek the approval or condescension of others, you will have a negative effect on your self-esteem. It is crucial that you approve of yourself, so that you feel no reproach for not taking care of yourself.

Change of station. Do not compare yourself with others; if they are ripe with good health or with illness, that is their problem, do not say, "He or she is worse than me, it consoles me". Everyone should evaluate themselves independently, without outside reference points, this will allow you to draw your own standards and avoid giving the power of decision to others.

Toxic phrases to erase from your mind

There are phrases in the minds of those who blame themselves for being sick; they must be suppressed to reduce their effect:

- "I should have known better!"
- "It's all my fault!"
- "I blew it!"
- "If only I could turn back time."

Answer as sincerely as possible and start from there to heal:

- What do you want to be cured of?
- What is your attitude towards an illness?
- What are your most frequent emotional states in a day?
- What is your share of responsibility for being sick?
- What is the price you have paid for being sick?

- How has your illness affected those around you?
- What possible lesson does being sick teach you?
- Have you already made the decision to be cured?
- Where do you need to start?

Consciously draw up an action plan (including a plan B) that specifies hour by hour what you must do to get cured of that illness that has left you lying in bed in pain, that incapacitates you or that leaves you half-happy. You will need small supports such as clocks, calendars, agendas, anything you can think of to keep in mind what you are going to do and the times to carry it out. These types of reminders help you to keep in mind that soon you will find the relief you have been seeking for so long. Get out of your comfort zone or normalise your illness.

At first it will be scary, especially if you have been sick for years and trying, trying, trying. But the first step is crucial to gain ground, do not do it all at once and without a parachute, but in small doses, with small changes like choosing a very healthy habit: running, yoga, quitting smoking, saying "no" to that person. There are no pills or magic formulas, just focus with vision on what you want to attract into your life, what you have tried to make of yourself as a healthy person. Focus on your strengths (not your limitations), what makes you strong physically, emotionally and spiritually. Execute your plan now! But do it with the three P's: patience, persistence and perseverance.

Chapter 9
What if it doesn't work?

This is the typical question that many people ask themselves when making crucial decisions, whether it is about a profession, a change of job, a partner, a risk, a surgery, a health option outside the known where certain drugs generate "security", comfort... And there are so many times that we prefer "better the devil we know than the devil we don't", because this has always calmed our symptoms, it puts the pain to sleep for eight hours. The known offers us protection, the comfort zone is like that rickety armchair that we do not want to change because we feel safe there, leaving it generates anxiety and a certain fear, we prefer that place. We go out to try or because we are encouraged, but we return immediately, there we are safe, on automatic pilot where we already normalise being sick.

However, opening the door, leaning out and little by little leaving that zone, will be an antidote for your physical and emotional health. You will expand your limits beyond the known, you will know other solutions, other doors, you may stumble, but the benefits will be greater. You will encounter changing environments that will test you to bring out your resilience and strengths. At first, no doubt, it will be frightening, especially if you have been parked there for years, between pain and hope, but the first step is crucial to gain ground, do not do it suddenly and without parachute, but in small doses, small changes such as thoroughly researching that therapy about which many speak of its excellent results. Some doctors will tell you that it is not an alternative, although this is due to lack of information on the subject, so I recommend that you approach a specialist who masters it.

What if it does work?

The question acts as a consequence. Fear is a natural instinct that keeps us alert, the dilemma is that for many this emotion functions as the starting point in decision making - and this fear includes questions such as:

- What if it doesn't work?
- What if it gets complicated?
- What if I don't solve anything?
- What if I get cancer?
- What if I undergo stem cell treatment but my life changes?

Of course it will change, but for the better, that's for sure! Many of our problems begin with the familiar question, "What if?" "What if I die from this diabetes?" Of course it can happen if you don't take care of the problem. Confidence is a valuable attribute, it helps us to feel good, to get better results in what we do.

To have confidence is to dare, it is to make decisions and take on challenges, it is to go for a consultation and undergo that vital stem cell therapy. It is a kind of attitude that allows you to have certainty and confidence to opt for that treatment. Ask yourself, "What is the worst that can happen to me if I undergo this regenerative medicine?". Consider the worst-case scenario, it will certainly not be so tragic if you visualise an intervention with a specialist with professional experience in the field; there will be great benefit, since there will no longer be pain or risk of the disease growing.

The only thing you can be one hundred percent sure of is that if you don't take care of your health right now, later on you will pay the cost. Many patients stay on the bottom rung because they begin to look at obstacles that translate into, "I guess I had better not take a chance on this". They begin to think the worst, like that inner terrorist that kicks in to scream, "Don't do it!"

Gag him by reminding him of the times you dared to go on that strict diet and succeeded; we know that diets don't always work, but it did work and that's what I propose, lifestyle changes. Remember the times you took the risk to make another decision, and everything turned out well. Most of us, when faced with these types of scenarios, turn on autopilot or react instinctively, and it's even normal, since we intend to change what we have always done. "I won't be able to", five words with a determining mental effect, which the subconscious picks up on to proceed, override it so it doesn't sabotage you!

Focus on your strengths (not your limitations), what makes you strong physically, emotionally and spiritually. Log into your mental computer and start changing old patterns that have you in pain and worry. If you could run back in time, what would you like to change about your life? Surely you'd change sleepless nights, binge eating, bad eating, sedentary lifestyle, alcohol and so many drugs with side effects?

Will you give stem cells the benefit of the doubt? What do you want to change, when and how will you do it? Think resourcefully and make a log that includes a crucial change of toxic foods and habits. Choose that goal, realistic and specific, be focused, concrete what you work for, it's time to make those decisions that will bring you health, holistic wellness, a life of dignity!

Assertiveness enables you to face many daily problems and your health is one of them, not the expectations of others, but your own. Remember; what matters most is not how others see you or what they think of your decisions, it is how you see, value and take care of yourself. From today try to love yourself more, give yourself the benefit of the doubt and everything else will follow: physical, emotional, spiritual and financial health. With enough self-love you will be able to glimpse paths full of health. Everything moves in a single path. If you love yourself enough...

Chapter 10
5 valuable reasons why you should love your body!

Have you ever thought about what happens as soon as you bite into a delicious apple? Well, an immense machinery inside you starts moving its cogs to send signals, chewing, tasting, digesting, absorbing nutrients and storing energy. Loving your body means giving value to every function it generates for you, so that you continue living, enjoying, loving.

Reason # 1 Because it beats for you

Heart

Without a doubt, your heart is the engine that keeps you alive, thanks to its strong valves that push thousands of gallons of oxygenated blood through every corner of your body every day, enough to fill 200 large pipes, with a capacity of 8000 gallons each. One of the best ways to thank it is to take care of it with exercise, healthy eating and, above all, GOOD FEELINGS.

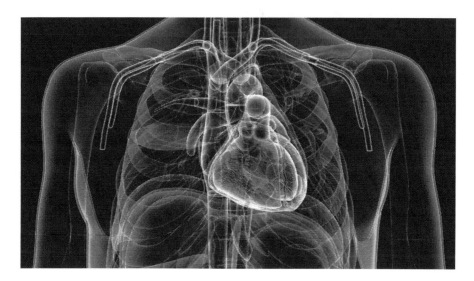

Be happy, love, be grateful, give a hand, smile... And if your heart is happy, you are happy too, and alive. Being happy is an attitude towards life, each one decides what, how, why, where and how much to be filled. Life is like riding a bicycle; you fall only if you stop pedalling. Be happy for you, for what you are, so that this quality does not depend too much on what you have or what you are given.

Love it a little +

- It takes 90 seconds for blood to transit through the circulatory system.
- In one day, a healthy person's heart beats 103,680 times.
- Men have more blood than women (5.5 litres for men, 3.3 litres for women), and their blood is also richer in red blood cells. Each cubic centimetre contains 4.6 to 6.3 million red blood cells, while women's blood contains only 4.2 to 5 million.

Reason # 2 Because it supports you

Every day, that body that has been given to you makes you undertake a miraculous journey, expanding borders, overcoming obstacles - it goes, comes, goes up, goes down, jumps, etc. And all this, for the most part, thanks to your bones, screwed together in an amazing feat of body engineering. Your skeleton is made up of

206 rigid bones, and is a calcium structure that keeps you upright and forms a protective cage around your delicate internal organs. It protects you and leads you by the hand through your life. It is estimated that in a year a person takes about 5 million steps and each one contributes to the formation of our bones. Every action is followed by a reaction. Exercise keeps them healthy; a sedentary lifestyle weakens them.

Reason # 3 Because it protects you

The skin is the largest organ of your body, weighing about four kilograms and covering an area of almost two square metres. Millions of cells are arranged in the form of tightly packed tiles, making your body waterproof.

The wonderful thing about it is that it is completely renewed every month, so you're protected behind it. But apart from being a formidable barrier, the skin has to be flexible and sometimes extremely adaptable. For example, pregnant women produce a hormone that allows their skin to stretch to twice its size. Take care of it, moisturise it and keep it protected from UV rays.

Love it a little +

Your skin is also an active organ with a vital function - keeping you cool. When you exercise intensely, your muscles generate enough heat to boil several cups of coffee. This heat would easily damage your internal organs unless your body did something about it. Just beneath its surface are three million temperature-controlling units, sweat glands, each a metre-long, fluid-filled, coiled tube. As your body heats up, the tube contracts and causes beads of sweat to rise to the surface of the skin. As the sweat evaporates, it reduces body heat and cools the body.

Reason # 4 Because it feeds you

In essence your stomach is a biological blender, its lining covered with delicate folds that allow it to expand with each bite. Inside, a mixture of enzymes and hydrochloric acid begins to digest the food. Corrosive digestive acids are so powerful that they can pierce metal. However, they cannot pass through the intestinal wall, which is protected by an excellent mucosa. In addition, the cells are renewed at the rate of half a million per minute and the entire wall changes every three days.

Love it a little +

One reason to love it more is that, without a doubt, you can taste every piece of food you put in your mouth. The surface of the tongue has 9,000 chemical detectors. These taste buds require just a few molecules of substance to find whether something is sweet, salty, bitter or sour. And we can only taste these four sensations. All the flavours that our taste buds perceive are only combinations of these. What a delight!

Reason #5 Because it thinks for you

The most powerful computer in your body is your brain. Just to comb your hair, it has to send millions of nerve signals to your hand every second, while simultaneously co-ordinating a multitude of different bodily functions. These stimuli are transmitted at a speed of over 400 km/h. The brain is able to do this because it processes information in a unique way. It divides the work among its 100 trillion cells, and at the same time each one communicates with ten thousand others through a network of electrical connections - incredible, isn't it? And not only that, but behind everything you do there is an army of microscopic machines, responsible not only for brain actions, but for all bodily processes: 50 trillion cells.

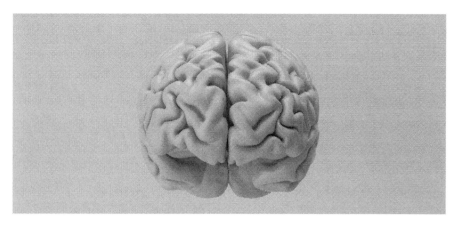

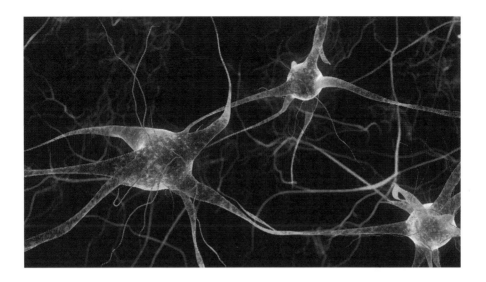

Your wonderful body is cells. Inside each one there are hundreds of electrical stations, protein factories, recycling plants and transportation highways. Thanks to all this gear, you move, think, do, create, have fun, fall in love. The human being has an average of 100 million neurons. For its proper functioning, keep it in shape by exercising, reading and eating healthily.

Love it a little +

Your mind is capable of experiencing all kinds of situations through your senses. The sense of touch uses 5 million sensory receptors that live inside the skin.

- Women have more connections between their cerebral hemispheres, which makes them more skilled in the mastery of emotions, while men have a greater capacity for abstraction and calculation. In addition, the corpus callosum, the structure that joins the two halves, is larger and this better brain communication could explain the famous female intuition.

I remind you again, love your body every day, not only telling it, but giving it the best care you can. To do so is to get away from all those bad habits that so far have damaged it (alcohol, cigarettes, drugs, sugars, fats, sedentary lifestyle).

Love your body!

- It's the only body you have - the one that will be on your side for a long time.
- It belongs to you, it is the only thing you own, no one can take it away from you.
- It's like a "machine" that has internal organs that do amazing things. Thank them for everything they do for you.
- It takes just as much energy for your body to displease you as it does for you to strive to feel magnificent. Just reverse your thoughts.
- No doubt your body loves you, don't make it feel lonely and sad.
- Your body dances with you, eats next to you, lathers you up, helps you tie your shoelaces and takes care of your sleep.

If he loves you and would even give his life for you, do the same - go now to heal that body that is damaged, broken, pierced, beaten and in so much pain.

Chapter 11
You want to heal... but do you sabotage yourself?

On the road to healing, doubts, pending issues, work, "lack of time" arise in the minds of those who desire it, allowing the disease to progress and grow like that snowball that becomes an avalanche of obstacles, obstacles and excuses that detonate a damaged gallbladder, a sick liver, a pancreas or a deficient kidney. There are as many excuses pending as grains of sand in the world and more if we go around and around: "Maybe tomorrow", "I'll see my times". These are excuses that in essence contain feelings of obstruction, fear, anguish, and uncertainty.

Self-sabotage results when you want to take care of that pending health issue that you have been putting off day after day, month after month, but when you continue with it, without realising it, you put your foot down looking for obstacles, inconveniences, knots and even negative and defeatist thoughts about yourself: "What if my tests give bad news?" Fear and indecision become rational arguments. People protect themselves; it is safer to put your foot in your mouth to justify your lack of determination, than to pretend. You simply do not expose yourself to bad news; it's a defense mechanism to avoid receiving an unfavourable prognosis, you prefer to continue carrying that pain, that discomfort, that intestinal discomfort, than to take action on the issue.

Unfortunately, we Mexicans are not so dedicated to a culture of prevention but might be better if I told you that you can beat cancer ten years before it appears, avoid a heart attack or banish the dreaded osteoporosis just by having a medical check-up every 6 months or at least annually. Ideally, women and men should regularly go to the doctor.

Self-sabotage is the shadow of all those circumstances that each individual does not assimilate or accept, beliefs rooted in failures, fears and oppressions that perhaps come from childhood and adolescence and that were never located or confronted, have taken root in the memory, and appear with power and conviction in specific situations of great responsibility, of decision making. Self-sabotage includes many unconscious, unnoticed acts that people create just at the moment they are about to start, continue or finish that treatment, or a major change in their lives.

Chapter 12
Maybe you just lack a little motivation

The spark that ignites your warrior spirit is called MOTIVATION, not only in the area of health, prevention, healthy eating, exercise, or a path of good living habits, but in all spheres of your life. It is an inner light that gives you the ability to focus your physical and emotional effort to achieve your goals; to heal and progress. It is that terrible desire to start, continue and finish a challenge, whether it is a wellness, work, family, professional or personal challenge. We all have that light of motivation, to fight and face bad news, but sometimes it goes out when we get frustrated or willpower weakens when you have tried again and again without luck.

Motivate yourself, go to your clinic, talk to your doctor, generate that synergy where you are the favoured one. For that inner gasoline to emerge, you need to find your own motivations, those of others are of no use to you. You need personal motivations for you to lift the towel and continue your treatment with good results. How do I do it? First connect with yourself; stop and reflect on everything that gives you strength, that paints your days and that gives your existence that extra vitality, on what makes you jump out of bed when it is still dark, raining, or very cold. Once you have done this exercise, try to include that energy in your daily routine, to accompany and encourage you constantly. Don›t leave pending tasks in little pieces of paper that get old because of time; do it here and now!

If one of your dreams is to climb Everest and you have a condition in your lungs from birth which apparently prevents you from doing so (of course, effort and dedication are key!) there will be avalanches and obstacles, as in any challenge in your life, but having a stem cell treatment can help those chemoreceptors in the pulmonary vessels to adapt better at altitude. Regenerative medicine may be the key to keeping your dreams from fading away. This is just one case, there are others, where your apparent inherited limitations condition you to realise your dreams. Do not be discouraged! Science has advanced and is advancing by leaps and bounds, anything is possible!

Everest

So go ahead, start that treatment, go to your doctor, start that diet, and achieve that goal you have set for yourself. Design a plan of attack with everything and a plan B, trust in yourself and do not leave it pending on the road. Life accumulates these extensions in thoughts as toxic as: "If I had...", imagining what would have happened by changing any decision made yesterday only causes dissatisfaction and doesn't change anything. So, from now on, replace these thoughts with "I will" and you will see what I am talking about. Take charge of your life, decide and act. Don't just read for the sake of reading. Decide to change with time, focus and alternate routes. Once you have made your choice, feel confident and trust that it is the best one.

Chapter 13
Medical empathy

It is under these themes of emotional strength that I put aside some of my scientific knowledge to generate an effective synergy between my patients and myself; today it is crucial that physicians lower their ego three notches to create empathy with those who approach them. In my case, it is in the consultation room that I carry out the therapeutic work using human warmth to build trust, security and greater mutual closeness.

I always listen with great attention to the pain and suffering of patients, including family, work and personal venting. It is like a catharsis that lets flow all those adverse emotions that compress us daily: lack of courage, anguish, frustration, discouragement, sadness and despair for not finding solutions, for knocking and knocking on doors without results. Listening, rather than hearing, that is the essence of a good doctor. It creates confidence in patients to be listened to, and they almost always respond well to those who do. Listening is a tremendously powerful tool in getting my patients to see the world as I see it. If all doctors, in all specialties, would learn this art of listening, believe me they would achieve incredible empathy and they would give it right back to you the patient.

I also talk to my patients in a personal way, one on one, without my ego and qualifications getting in the way. I explain in plain language and great detail about the disease they are suffering from, why they got there and along the way I offer guidelines to follow; health options that include stem cells and their good healing performance. Then I explain and emphasise what the procedure is about; what it is, what it is like and what will happen when it is applied, including the probable side effects, such as fatigue for the first 24 hours. Finally, I clear up doubts, because, until the patient enthusiastically says "Let's do it!" the procedure does not begin. And after a health protocol, under strict safety conditions, I apply the appropriate dose. So much for my 50% of the work.

The rest depends totally on the patients themselves, each one must do his or her bit for good health to crystallise. In other words, stem cell therapy produces satisfactory results as long as people also change their suffocating lifestyle. At this point, stop what you are doing, at least for a few minutes. Take a short walk to a nearby park, breathe, free yourself, come back and start your task again, you will see what I am writing about. For there to be a cure as such, feedback, work and synergy on both sides are required – this harmony soon yields good results.

It hurts me that it hurts you...

Emotional empathy is a very popular feeling nowadays, to put oneself in another person's shoes, to understand their state of mind. This perception is possible due to mirror cells, which are important for the brain, as they give us the guideline to learn, associate and see ourselves reflected in other individuals. Basically, this is what we use during the first years of life and we learn by imitation.

Did you know that...?

- If your waist is more than 80 cm, your risk of type 2 diabetes increases by 50% and by 20% for heart disease. Measure yourself!
- These tests are useful to detect possible tumours, clots or obstructions in blood vessels, which can cause a cerebrovascular infarction. A check-up is vital, especially if you have a family history of these.
- How is your blood pressure? The normal range is between 120-80 (millimetres of mercury) for men and 110-70 for women. Both high blood pressure and low blood pressure are dangerous. According to health institutions in Mexico, 1 out of every 4 Mexicans is hypertensive, and the number will soon be higher. A person can die if his or her blood pressure goes through the roof or drops to the floor.
- A million bacteria fit on the head of a pin and if you grab it, you could die... But if your immune system is a fortified bunker, it won›t get past the pinprick. Your defense system is a natural army of cells like T cells, which keeps viruses, bacteria, and fungi at bay; they detect, block, destroy and swallow invaders... Sleepless nights, stress, poor diet, pollution and sudden

changes in temperature weaken your defences so that any disease can harm you. Strengthen them! Eat well, exercise, and stay away from bad habits.

Chapter 14
Good life habits

Poor eating habits lead to multiple diseases such as diabetes, obesity, hypertension, heart attacks, cardiovascular problems and cancers.

Surely at this moment something triggered in you by what you are suffering or by watching the suffering of a loved one. This can be pain, suffering, helplessness, physical, emotional, financial and even spiritual wear and tear. Sad and pitiful is the way of life that most have followed. It is commonplace that in the morning people run and go to work with nothing in their stomachs. After an eight-hour fast, the fuel level in the blood drops, and the consequences range from dizziness and fatigue, to an ulcer or gastritis, heart disease, high cholesterol and chronic degenerative diseases. In addition, it has been proven that diabetes and obesity are closely related to skipping the first meal of the day.

Prolonged fasting causes your metabolism to work slowly, and during the course of the day you begin to «stock up» on junk food. This in turn causes the body to adapt to larger reserves than it should, causing the body to become overweight and obese. It is clear that there are new trends which propose intermittent fasting, which may work for many but not for everyone - knowing which food to break this fast with is the key to its success.

It is urgent that you start now to look at where you are and where you are going. Making a stop in your existential pursuits is your first obligation. Stem cells, an optimal and accurate treatment, will bring you integral health and wellness as long as you discipline your habits to make changes that include eating well (and at the right time), not staying up late, reducing the consumption of cigarettes and alcohol, getting enough sleep and exercising, so if you are a healthy person, stem cells will act as an anti-ageing device.

If only you knew the great benefits of simply walking. Walking is a popular and very economical activity. It is free, safe and effective for your health. Without changing the rhythm of life too much, you just have to get off at an earlier train stataion or park farther away than normal and walk the remaining distance to the office or wherever you have to go to at a normal pace.

6 benefits of walking

- Less fat. If you carry extra pounds, going outside and walking for 30 minutes at a speed that matches your heart rate will certainly help you burn fat, stay in shape and feel full of energy during the day.
- Strong muscles and bones. Walking as a life routine forces the body to support its own weight, and additionally your bones will achieve greater density and resistance to keep you upright for many years. In other words, you reduce the risk of osteoporosis by up to 50%.
- Useful thoughts. The brain commands, remembers, thinks, memorises, reasons, discerns, feels. It never rests. It works day and night like a powerful biocomputer, so it needs enough food to fuel it for those 24 hours. A walk around the block will help to oxygenate it, clear your mind and foster your creativity.
- Strong lungs. When walking, twice as much air enters your body, twelve litres instead of the usual six, which multiplies the lung capacity by four; the oxygenation of the blood and the whole body will be in perfect condition.
- Improve your blood circulation. Poor circulation is a common symptom of high blood pressure. Regular walking gets your blood flowing and reaches all the organs in your body.
- Heart of steel. Undoubtedly, this organ is the one that will benefit the most, since the oxygen that enters it is ideal to help prevent heart attacks or cardiovascular problems.

Did you know that...?

Centuries ago, the British physicist Thomas Sydenham said: "The arrival of a clown in a village would do more for the health of all its inhabitants than twenty donkeys loaded with drugs." Years later, in 1971, the physician Hunter Patch Adams ("Doctor Joy"), revolutionised hospitals in the United States by proposing laughter therapy to improve the health of the sick. It would be great if, when we visit the doctor, he would prescribe us a laughter pill each day.

When we laugh heartily, twice as much air enters, twelve litres instead of the usual six, which increases lung capacity and blood oxygenation fourfold. All your cells will be in perfect condition. In addition, our body undergoes a series of involuntary changes that we do not perceive. Four hundred muscles relax, digestion improves, blood pressure levels out, notably reducing the risk of heart attacks - incredible, isn't it?

Chapter 15
Obesity, the bread that generates many diseases

Not long ago skinny people were viewed in a negative way, they were thought to be sick and weak. A popular myth has been in relating weighing 100 kilos with happiness, when the reality is far from it. This is a sad state of affairs and we doctors see it every day in our patients. And just like this belief, there are so many others that become myths or lies along the way, here are some:

Lie # 1: Fat, but happy

A chubby child of yesteryear was a "healthy and happy" one. Parents and grandparents, not so long ago, instilled that maxim in us. That kind of health and happiness leads to serious heart disease. Those who suffer from excess weight generally suffer from anxiety and become depressed. Excess fat lowers the levels of dopamine (which controls depression).

Sad reality. Due to the high levels of overweight and obesity that prevail worldwide, but especially in Mexico, colon, stomach, breast, endometrial, ovarian and prostate cancers have increased, warn several health institutions. And accordingly, 150,000 new cases of cancer are registered in Mexico each year, hence the urgency of taking action on the matter. There is no doubt that stem cells are the aces of good results.

Numbers:
- A Mexican consumes 163 litres of sugar-sweetened beverages per year.
- 33 out of every 100 Mexican adults are overweight.
- One in four children between 5 and 11 years of age is overweight.
- Eighty-five percent of obese children are also obese as adults.
- Diabetes is a consequence, and we are also #1 in patients: one out of every 6 Mexicans suffers from it. According to data from the Ministry of Health, 70,000 people die every year due to complications from this cause.

Did you know that…?

Obesity can be caused by damage to the thyroid gland, which is responsible for generating hormones for proper metabolic functioning.

How do you know if you are obese? Be careful, being overweight is not the same as being obese. The former is an excess of body weight in relation to height, while the latter is a disease characterised by excess body fat. Calculate your body mass index (BMI):

BMI= Weight in kilos divided by height squared. Example: if you are 1.60 m tall and weigh 68 kilos, you must divide 68 by your height squared (1.60 x 1.60) = 2.56. Check it out.
- Underweight = 18.5
- Normal weight = 18.5 to 24.9

- Overweight = 25 to 29.9
- Obesity grade 1 = 30 to 34.9
- Obesity grade 2 = 35 to 39.9
- Morbid obesity = 40

Lie # 2: I'm going on a diet in January...

Generally, diets do not work for an obese person. These are undertaken as a punishment, against the will and, although the internal emotions say otherwise, the person forces himself not to eat. When he/she can't take it any more, he/she binge eats, gains weight and then feels guilt, reproach, self-loathing. Then he/she punishes him/herself and goes back to the diet, now stricter. Thus he/she goes round in circles that can lead from illness to death.

Lie # 3: Oh, well, I'll just take care of it!

Treating this disorder is difficult because it is chronic. First, the patient has to assimilate that he/she suffers from it and to what degree. Then, the causes are assessed and the triggers, such as sedentary lifestyle, binge eating, lack of exercise, etc., are modified as we go along. Plan B consists of resorting to a medical treatment that may include doses of stem cells, since these accelerate the metabolism. It is all a treatment that begins in the mind; remember, you are what you think, that amazing law of life that brings you more of the same. If you want to be cured, start by being aware of what you suffer from, and from there approach alternatives different from the conventional ones, those that only cure the symptom.

Chapter 16
Eat healthily... and nourish your cells

All human history testifies that, ever since Eve's morsel, man's bliss has depended on food.
Byron

Undoubtedly, food plays an important role for all your cells to perform their corresponding tasks 24/7. If your plate and your cup are empty, and you fill them with fried food and tortillas, obviously your cells will be like that: sick, slow, idle, disoriented. From the bad food you put in your mouth, many ailments such as hypertension (also called "silent death") arise, as with diets high in sodium, soft drinks, sugars, fats, alcohol.

Undoubtedly, food plays an important role for all your cells to perform their corresponding tasks 24/7. If your plate and your cup are empty, and you fill them with fried food and tortillas, obviously your cells will be like that: sick, slow, idle, disoriented. From the bad food you put in your mouth, many ailments such as hypertension (also called "silent death") arise, as with diets high in sodium, soft drinks, sugars, fats, alcohol.

Vascular System

A diet low in sodium and abundant in fruits, vegetables, whole grains and low-fat dairy products is necessary for the heart to circulate blood properly. Don't forget to consume enough vitamin C naturally (75 mg per day for women, 90 mg for men); this is easy to obtain through fruits and vegetables, for example, a lemon usually contains 40 to 53 mg. Your diet is crucial!

A few weeks ago the WHO ranked Mexico as one of the sickest countries on the planet, physically and emotionally. We are already # 1 in diabetes, according to the SS, there are more than 12 million registered cases and this number is not accurate, because there are many people who do not even know they suffer from it.

Diabetes Testimony

We are one step behind the United States with regard to obesity so the future points to an obese Mexico and this, in turn, generates many cardiovascular and image problems. Smoking, which is already commonplace in the streets, is very close to it. According to the National Addictions Survey of the Ministry of Health, smoking is the third leading cause of death in Mexico and today women are the ones who smoke the most. Not only that, but hospitals, clinics and health centres are already overflowing with hypertensive, infarcted and malnourished patients, etc. According to the world institution, stress and depression are one of the main causes of work absenteeism.

It is urgent that you take action if your health deteriorates, first of all, remember to consult a doctor, do not skip it or leave it for another day, your health is important. Give your cells a break to regenerate, repair and restore. And if as a treat you eat a hamburger, good for you, but accompany it with a refreshing cold water and a salad. Remember, it shouldn't be every day. Do not condemn or demonise certain foods, as long as you eat them occasionally and in moderation.

Chapter 17
6 table tips you shouldn't overlook

1. Balance your ingoings and outgoings. One way to achieve synchrony between what you take in and what you spend in energy is to have a diet that includes a whole range of foods.
2. Integrating is the route; balancing is the key. Remember that all excess is bad. There is no such thing as bad food, everything can be included in a healthy diet plan. It is true that so-called fast food is usually rich in fat, calories, cholesterol, sugar and sodium, however, if it is eaten occasionally, it will not cause you problems.
3. Let your food be served as naturally as possible. Eat fruits, vegetables and seeds, in their totally natural state, fresh from the earth, washed and properly disinfected then eaten, to take advantage of the sap, which is the vital energy of plants. With this you get all that natural energy.
4. Occasionally consume frozen, canned, packaged, and processed foods, as these contain preservatives such as sodium, colourings and additives to make them last for years.
5. Eat all types of meat in moderation, they are sources of protein, zinc, iron and selenium.
6. Learn to read the nutritional information of the products to be purchased. Look for foods high in protein, low in carbohydrates, high in fibre, low in sodium. Eat healthy food, suitable for human consumption, along with plenty of water, exercise, and enough sleep, which will detonate cellular energy, strength and regeneration in every geographical corner of your body, internally and externally. All your organs, arteries, blood vessels and systems will thank you.

Chapter 18
"Today is better than yesterday, tomorrow is better than today."

Nowadays there are many people concerned about their health, in search of what they considered lost after many years of being stuck in an "incurable" condition. They are people like you and me who long to feel well, no longer bedridden, disabled, in pain and permanently suffering, so much so that they have already accepted it as a daily norm in their lives. These are patients who, after knocking on countless doors, aspire to overcome the diagnosed disease, which to date has only been treated with painkillers every 8 hours, numbing the symptom, but not getting to the root cause. They are people who dream of improving their quality of life, soon and without excuses. They wish to get out of this vicious circle that immobilises and keeps them away from work, family and friends. They long to sleep so many hours, in a restful sleep free of torment, to rest, eat, live and enjoy what they have not been able to due to the disease and the medications with their side effects.

Stem cells are here to modify somewhat the concepts of disease and cure; they offer encouragement, hope, certainty and good news in various areas of health, including sexual health. All these things and more are what your cells can generate naturally, automatically because they already have it in their DNA, when they are healthy or you stimulate them in an efficient and correct way. It is a pity that the function of these internal elements nowadays has been altered, reduced by so many bad habits of life such as bad diet, sedentary lifestyle, stressful work, late nights and the consumption of toxic substances. Therefore, good health begins by changing this harmful lifestyle. Your health depends on you, remember that.

Not long ago I read about the kaizen philosophy and I was attracted by a short but powerful phrase: "<u>Today is better than yesterday, tomorrow is better than today</u>". These words apply when illness affects your body, as you depend on your willpower, nutrition, exercise and effort to be better every day. The stem cell treatment emerges as a mechanism of strength to continue, not to procrastinate and leave it halfway.

If you are concerned about that condition, this maxim encourages you to solve it while immersing yourself in small daily changes and constant improvement in the different areas of your daily life. The kaizen philosophy shows us that we should not let a day go by without doing something new or different in our wellness log, something that keeps us healthy such as exercise, a balanced diet, and sleeping well.

Every journey is made up of small steps. It is about breaking down the goal you have in mind into small tasks, taking one step at a time, one focused, concentrated and continuous step. This philosophy affirms that perseverance and getting up again after an ordeal is the soil where achievements germinate. Be resilient, choose how you want this sunny day to go, what happens to you today, how you want to feel physically, emotionally and spiritually. The mind is a powerful computer and receiver of what your lips emit, it will focus on accommodating various spheres of your life, do not forget that your health is one of these and one of the main ones.

One step at a time means not feeling overwhelmed to meet a great challenge, which is to banish once and for all that disease that has even made you bedridden. Make a plan to confront it, divide it into units so small that they do not sabotage you or make it impossible for you to carry them out. Do you want to exercise? Do it for 10 minutes and add it up in time. You will create a productive habit, the foundation of any transformation. These small changes, which you must persevere with, will have an impact on your life in a short time like the snowball that starts small and in the end is an avalanche of good news ... One of them, no doubt, will be to flaunt in your agenda a stem cell treatment.

Successful Cases

Index

Dedication	5
Introduction	7
Chapter 1	
Stem cells, a myth-busting science that creates hope	11
What comes to mind when you hear or read "stem cells"?	12
Chapter 2	
Stem cells: scanning,	
detecting and repairing	15
Placenta cells from another person inside my organism?	16
Preparing the patient	18
Chapter 3	
How do stem cells work??	21
Side effects?	22
Chapter 4	
10 curiosities about stem cells	25
Chapter 5	
Autoimmune diseases and how to treat them	29
When my own body attacks me... and generates autoimmune diseases	30
Dare to change your life!	31
Chapter 6	
Between your body and emotions	33

Chapter 7
We are mind/energy — 37
We are mind — 37

Chapter 8
Do you feel guilty about being sick? — 41
Toxic phrases to erase from your mind — 42

Chapter 9
What if it doesn't work? — 45
What if it does work? — 46

Chapter 10
5 valuable reasons why you should love your body! — 49
Reason # 1 Because it beats for you — 49
Reason # 2 Because it supports you — 50
Reason # 3 Because it protects you — 51
Reason # 4 Because it feeds you — 51
Reason #5 Because it thinks for you — 52
Love your body! — 54

Chapter 11
You want to heal... but do you sabotage yourself? — 55

Chapter 12
Maybe you just lack a little motivation — 57

Chapter 13
Medical empathy — 59
It hurts me that it hurts you... — 60

Chapter 14
Good life habits — 63
6 benefits of walking — 64

Chapter 15
Obesity, the bread that generates many diseases   **67**
    Lie # 1: Fat, but happy   67
    Lie # 2: I'm going on a diet in January...   69
    Lie # 3: Oh, well, I'll just take care of it!   69

Chapter 16
Eat healthily... and nourish your cells   **71**

Chapter 17
6 table tips you shouldn't overlook   **75**

Chapter 18
"Today is better than yesterday, tomorrow is better than today."   **77**

Made in the USA
Las Vegas, NV
20 December 2024

d446f6fd-b42e-41a0-9617-12f3f112d685R01